THE LAST TEMPTATION of MAX GOLD

by MAX GOLD

THE LAST TEMPTATION of MAX GOLD

Written December 2013. Copyright © 2014 Max Gold

All rights reserved. The onus is on the reader to attain a fully detailed and comprehensive cognizance of all the rights. All of them. This book is sold subject to the condition that it shall, by way of trade or otherwise, be lent, resold, hired out, or otherwise circulated without the publisher's prior consent in any form of binding or cover other that in which it is published and without a similar condition including this condition being imposed on the subsequent purchaser. The scanning, uploading and distribution of this book via the Internet or via any other means can be done without a greenlight from the publisher.

Please do, however, purchase only authorized editions of people's creative labor. Piracy doesn't help people who make the books your read, the books you may not read that become the movies you watch, the movies you watch and the music you listen to... make money. If people can't make money, they won't make the good shit you enjoy; or at least there will be a lot less of it for you to enjoy.

Piracy is the reason people won't take risks investing in questionable musicians, filmmakers and writers. Risk aversion leads to the reign of pop stars, reality stars and every other form of mediocre that you're bombarded with on a daily basis. Your support of other artists rights is apreciated.

Max uses Microsoft Word, Quark XPress 10.0 and Uni-Ball pens.

LCCCN: 69-6969-69696

Printed in a factory.

Also by MAX GOLD:

"DON'T JUDGE a DICK by ITS FORESKIN"
God, Life & Revolution

dont-judge-a-dick-by-its-foreskin.com

~

"THE LAST PROPHET"*

the-last-prophet.com

Both available in hardcover through amazon.com

*also downloadable free of charge at the-last-prophet.com

THANK YOU

Luke (for putting up with the constant bullshit), Lesley (for the drive to Philly), Emma (for the ears)...

& Leslie.

~

DAY I...... 1

NIGHT I....... 6

DAY II...... 8

NIGHT II...... 16

DAY III...... 19

NIGHT III...... 26

DAY IIII...... 31

NIGHT IIII...... 37

~

Day I

I'm on a $21,000 vacation for four days and four nights, lying in a cabana, surrounded by nothing but white sand and ocean, with a cold beer in one hand and a cigarette in the other. Only a blowjob from the towel girl could make life better at this moment.

And then a couple in the adjacent cabana start going on about their Aunt's recent diagnosis with ovarian cancer; the shock, the bills, the kids, the odds, if her promiscuity had anything to do with it and, all in all, how screwed she is and how awful it has to be to face death.

Staring up at the sun, it hits me: No one gets in alive… you were dead before you got here.

I call the server over to my side. Taking 30 minutes to get a Corona is unreasonable at any time, let alone when it's desert temperature outside. So I ask for six on ice. Hector, as his badge suggests, says that he can't accommodate the request because no one has ever asked for "sees" before.

MAX GOLD: *(removing sunglasses and sitting up)* Well Hector, why don't we put the ass in class and serve them up on ice in a couple of champagne buckets?

HECTOR: Sees senor, sees.

MAX GOLD: *(palming Hector a $20 bill)* Thanks my friend.

Two minutes later half a case of Corona is delivered at my sandals in champagne buckets packed on ice in a lobster crate.

MAX GOLD: *(palming Hector another $20 bill)* Thanks my friend.

This is good. And then I drift into thought…

"It usually takes the diagnosis of a terminal illness to bring people into a conscious cognizance of their mortality. They get diagnosed and then it's all about them, the fundraising, the cure, the progress of treatments and so on. There is nothing wrong with this. It's human. Most people don't give a shit about a problem that doesn't affect them directly. Survival of the fittest...

People need to face death before they can appreciate life...

I've seen more people die, more people die on the inside from the death of loved ones and more people walking mortally wounded from the life stolen from them though any number of brutal experiences. I've known people who should be living and the fuck's who'd be better served in their place...

We are all diagnosed with death and shouldn't need to cough up blood at 5AM, feel chest pains when an escalator is out of order, or have a doctor tell us too many cells are growing in one place, to state the fact. The second you scream; death, it's a safe bet down the road...

If more people could see the ride we call life for what it is, a fleeting experience, the experience would be better. Wouldn't mean bills wouldn't have to be paid, obligations could be ignored, or consequences weighed insouciantly. It would just mean making the experience more meaningful, appreciated and fun...

MAX GOLD uncaps a beer and flicks the cap into the crate…

You wouldn't be afraid of death because you'd know death is your original (previous) state. Knowing this you'd be less afraid of this existence...

You'd laugh (on the inside) at unreasonable people and try and find the lesson they were teaching you...

You'd spend more time giving, unconditionally, to people you like and less to those you know deep down you don't...

You'd want to spend more time, even just minute exchanges, with friends, family and strangers...

You wouldn't get as angry. And when you would, you'd laugh at the absurdity and follow the path of least resistance...

You'd communicate better with people by just being more honest...

You'd have less ego because you'd have less to hide...

Making sure no one is looking, MAX GOLD urinates in between Cabanas...

You'd want to try new things...

You'd want to learn, read, listen and think more...

You'd actually appreciate a good meal and know what went into getting it in front of you and why you should be appreciative...

You wouldn't need a motivational speaker because you'd be able to hear your own voice...

You'd pause to yourself in the middle of having a blast and say... this is awesome, I'm lucky I'm not homeless and I wish homeless people weren't so...

You'd treat every woman you fuck like the sexual goddess they are and share your body and soul with them like it were the last time you could...

Not caring whether any one is looking, MAX GOLD urinates in between Cabanas...

You'd help people with your money if you could afford it...

You'd help people with your time if you could spare it...

You'd realize the work always gets done...

You'd find ecstasy in solitude...

You'd find solitude in a friendship...

You would drink from a glass of wine and feel history flowing through your body...

You wouldn't care about dying because you spent so much time being aware of living...

You'd write your own list – and follow through on it...

You are alive. And only once. Good luck and enjoy...

Don't be afraid of dying...

It's more of a trip home...

Just don't be an asshole before you leave the party."

MAX GOLD nods off and spills his beer on his chest, which wakes himself.

As the sun sets, I head inside to flake out in my room and get ready for dinner.

NIGHT I

After a rock star meal on a mountain top overlooking the ocean, I manage to engage in warm conversation with the table beside me, while a woman sitting at the restaurant bar takes notice.

By the time the bill is paid I have a drink left and the couple's limo has arrived to pick them up. We say our goodbyes, I politely decline the extended offer to sightsee with them the next day and then they make their way.

And then I head to the bar.

It's been my experience in life that dictates there is only one kind of woman who opts to sit beside a strange man in an empty bar. A hooker. And may god bless every one of them.

WOMAN WHO CHECKED ME OUT EARLIER: So what brings you here all alone?

MAX GOLD: Same as you I suppose. Just looking in life and trying to find some peace before heading back to the storm.

WOMAN WHO CHECKED ME OUT EARLIER: Are you into astrology or Rikki Healing because…

DAY II

I wake up surprisingly not hung. It's 11:11am.

WOMAN WHO CHECKED ME OUT EARLIER is beside me and it dawns on me that I either made it back to hers or I'm in a hotel room.

Quietly, I slide out of bed and head to the kitchen in search of coffee. After making a cup, I walk over to the coffee table to put my jeans on; laughing to myself as I slide them over my waist and feel the chaffing. I walk out onto the terrace and light a smoke and say "ya baby" to the sky. The words and harmony "Cause I'm a motherfucking P.I.M.P." by 50 Cent start streaming through my head.

For whatever reason, I drift off to think about Lauren. We had a masterfully fucked up relationship, but in spite of it all… I still think about her. I'd love to meet the one. Just the one that came without the baggage of a fully loaded 747… from Bombay en route to JFK.

Maybe one day I'll find the one.

Until then… fuck first, ask questions later.

As I try and slip past WOMAN WHO CHECKED ME OUT EARLIER, she stirs…

WOMAN WHO CHECKED ME OUT EARLIER: Hey honey.

MAX GOLD: *(Imitating Hank Kingsley)* Hey now!

WOMAN WHO CHECKED ME OUT EARLIER: What are you up to today?

MAX GOLD: *(thinking to himself)* She's a civilian… yes!

(placing a plastic card on her bedside table)

I'm at the Circle Beach Resort. Just say you're a guest of Max.

WOMAN WHO CHECKED ME OUT EARLIER: Ok. See you in a bit.

MAX GOLD: *(kissing her on her right ass cheek)* See you.

Back at the resort I managed to score a great front row cabana with an unobstructed ocean view. By noon I'm 4 Coronas deep and the book I'm reading about David vs. Goliath is getting interesting. Humanity loves rooting for the underdog. But humanity, like the favorite, often fails to realize the underdog always comes to play and has disadvantages that can be used to the opposite effect. After 8 Coronas I realize the book is just a recollection of highly entertaining stories, spliced with common sense.

I put the book down to take a piss and appreciate the beauty surrounding me… the waves crashing after their breaks… the birds floating aimlessly above… the steady shuffle of the clouds… the warm wind breezing through my shirt… the sun glowing down on everything… the sun turning black…

MAX GOLD: *(looking at the sun perplexed)* What the fuck?!

And then all around me becomes violently still. The waves stop crashing and become frozen in their break… the birds become suspended in midflight… the clouds stop moving… the wind stops breezing through my shirt and the sun starts to flicker. The whole earth goes completely black for an instant and then then daylight reemerges. Life around me, however, stays still.

I turn around to see a man dressed in a white linen suit lying in my cabana. He looks to be about 5' 2", heavily balding, overweight, with a ginger complexion.

MAN DRESSED IN A WHITE LINEN SUIT LYING IN MY CABANA: I can see why you forked out all that bread… this place is heaven.

MAX GOLD: Well, it was heaven until you showed up. Can I help you?

MAN DRESSED IN A WHITE LINEN SUIT LYING IN MY CABANA: Maybe.

MAX GOLD: *(feeling awkward)* Ok?

MAN DRESSED IN A WHITE LINEN SUIT LYING IN MY CABANA: *(biting his lip)* A… what would you call him… former associate of mine paid a visit to you a few years ago.

MAX GOLD: *(incredulous)* Wait a second… you're an archangel?

MAN DRESSED IN A WHITE LINEN SUIT LYING IN MY CABANA: Yes.

MAX GOLD: Then why are you so short? Satan was almost twice your size?

MAN DRESSED IN A WHITE LINEN SUIT LYING IN MY CABANA: *(caught off guard and visibly perturbed)* Well, that's because… *(cut off by Max Gold)*

MAX GOLD: Sure… so you're smaller for your kind. What's your name?

MAN DRESSED IN A WHITE LINEN SUIT LYING IN MY CABANA: god!

MAX GOLD: Holy shit… god is a fucking midget on his planet!

god: Watch it kid.

MAX GOLD: Or what?

god: Accidents have been known to happen.

MAX GOLD: Is that meant to put the fear of god into me? You know what else has been known to happen?

(god looks at MAX GOLD curiously)

Assholes stealing people's seats.

(MAX GOLD makes a dusting gesture)

Move over.

(god sits up and slides to the end of the cabana)

So what about my night with Satan?

god: I heard a SSA recording of your conversation the other day and it was quite troubling.

MAX GOLD: What recording? Is there an NSA in your dimension as well?

god: No, we have the SSA.

MAX GOLD: What the fuck is that?

god: The Supernatural Security Agency.

MAX GOLD: Really? Is that what you use to listen to prayers?

god: *(god laughs heartily)* I haven't listened to those annoying requests in a few hundred years. We use the agency to keep tabs on our satellite planets.

MAX GOLD: For control.

god: Bingo.

MAX GOLD: So you're short, bald, fat and controlling. What about my conversation was so troubling?

god: My concern is that if you go running around getting people to realize that religion is a crock of shit… that it's only purpose today can be to keep people and their spans of consciousness down… *(god pauses for thought)*… while urging them to form a fact based consensus on what they are, what the purpose to life is and what they can do as a species… well…

MAX GOLD: *(cutting god off)* Well what?

god: It was never our intension for you become a species of higher consciousness, capable of leaving this place and dimension, as one.

MAX GOLD: Why?

god: Because the universes have enough species in them and, quite frankly, they don't need an abomination joining them.

MAX GOLD: Wo, wo, wo… how are we an abomination?

god: Gee… let me think… You're all a bunch of fucking monkeys?!

MAX GOLD: It really is true isn't it?

god: Why do you think every human child is afraid of snakes?

MAX GOLD: Are they?

god: god ya. You can show any child on this planet a picture of a snake and they'll react to it with either disgust or fear.

MAX GOLD: No shit?

god: Oh ya. Even Inuit children and kids in Northern Russia lose their shit when first shown snakes. It's a remnant from the survival component of your genetic imprinting dating back to when you were monkeys.

(MAX GOLD cracks a Corona and offers one to god, but god declines)

MAX GOLD: Why are kids instinctively afraid of snakes, I mean they do look kind of creepy?

god: Many things common in nature are creepy looking, but you are all hard wired to fear snakes for one simple reason.

MAX GOLD: Being?

god: Monkeys spend a great deal of time in trees. They rely on trees for shelter, food and, most importantly, protection from predators.

MAX GOLD: But...

god: But while trees afforded the means for comfort and survival for monkeys, they left them vulnerable to one thing – snakes. Snakes in trees are a bitch for monkeys. Snakes are masters of camouflage, silent and can climb and hang onto things better than any mammal can. In fact, snakes are a problem for all animals even on the ground let alone up in a tree... and when a ground predator (or hazard) can kill you up in your tree-house... you have a problem. Over the course of hundreds of thousands of years the knowledge of this threat was literally passed on through your genes and still flows and passes through them to this day.

MAX GOLD: That makes sense and explains why Satan is portrayed as a snake in the bible.

god: Of course! Why do you think Satan is introduced as a snake slithering out of a tree? Shit, as the story goes, the snake is damned to crawl on its belly for the rest of its days because of what it persuades Adam to do with the tree.

MAX GOLD: *(rhetorically)* So when it comes to allegory, what better object to attribute your adversary to... than... the one thing we are instinctively programmed to fear?

god: *(smiling)* Brilliant, isn't it? Satan thought the story of the fall of man would teach people how to think for themselves... *(cut off by MAX GOLD)*

MAX GOLD: How so?

god: Well, the story shows that Adam and Eve defied me by choosing to eat the forbidden fruit; and as a direct result they realized truth.

MAX GOLD: What truth?

god: That I'm a liar.

The Serpent's Deception

...2. The woman said to the serpent, "From the fruit of the trees of the garden we may eat; 3. but from the fruit of the tree which is in the middle of the garden, god has said, 'You shall not eat from it or touch it, or you will die.'" 4. The serpent said to the woman, "You surely will not die!...

NIGHT II

MAX GOLD: How so?

god: I told Adam and Eve that if they ate, or even touched, the fruit in the garden of Eden, they would die.

MAX GOLD: What other truths did they discover?

god: That they were buck naked. That they have minds and a free will of their own. Most importantly though, they gained independence.

(god picks his nose and rubs his finger in between two cushions)

But there is a motherfucking snake in the motherfucking bible… and no one got the moral like Satan had hoped they would.

MAX GOLD: In the bible you're portrayed as, and come off as a pretty repressed, regressed and oppressing figure. Why is that?

god: *(smiling)* Because I am.

(MAX GOLD and god burst out into laughter. MAX GOLD opens another Corona and lights a smoke)

No, I am a dick, but you have to bear in mind that when we impart revelations, it's still men writing them down and fleshing them out. Men in the desert a few thousand years ago were less evolved in their logic, reasoning and social attitudes. My heaven, you could argue that case to this present day.

MAX GOLD: *(taking a drag)* But the bible has done a lot of damage. It's fostered harmful discord throughout humanity for centuries. It's impeded progress...

god: *(cutting MAX GOLD off)* Every religion and religious book came about as a much needed form of progress and social reform. You have to realize that while they are outdated today, they were lifesaving tools for your species at one time. They ushered in order in times of chaos. But of course, in the end, you fuck up everything. To be expected of over evolved monkeys I suppose.

MAX GOLD: You know I really resent all of this fucking monkey business. Your species had to start off as something. Everything does. Put a cock in it with all of your subjective prejudices.

god: Are you implying that your species is equal to mine?

MAX GOLD: I'm asserting it. We're just not as evolved yet. Everything is made of energy.

(MAX GOLD reaches for a bong. He packs the bowl, lights it and pulls a hit)

We're all...

(struggling to hold his breath, then exhaling and tilting his head away from the bong)

cut from the same cloth.

DAY III

god: I tell you what. If you can convince me of your true creator, then I'll know you have the potential that Satan believes all of you have.

MAX GOLD: Apart from energy, nothing.

god: Nothing? Whaddaya mean?

MAX GOLD: Nothing. Everything that is, ever was, and that will ever be... stems, ultimately, from nothing.

god: *(nodding)* Go on.

MAX GOLD: Anything that is in existence has to originate from something. And if you go back to the very first thing that ever was, well, what preceded it?

god: Nothing.

MAX GOLD: Exactly. But the reason humans can't grasp the concept of nothing (which is closely related to infinity in its difficulty to comprehend) is because of our perception of reality.

Humans see things as simply black and white, beginning and end. We have difficulty understanding that nothing, even if it is unconscious and devoid of energy - or devoid of having a beginning and ending... is still something.

god: Why do people have trouble with the concept of infinity?

MAX GOLD: Because we can't understand something not ending. Simply put, we have no frame of reference for something not existing or not ending. The experience and knowledge that our consciousness, and ego's, glean from our reality just can't show us how nothing (and infinity) are possible.

(MAX GOLD finishes his beer and pauses for a moment)

Infinity is a circle or a body of something in a constant state of expansion out into nothing.

god: But what gives birth to something... to energy?

MAX GOLD: Likely a spontaneous combustion of some kind that was brought about by nothing and is sustained by nothing, or itself (itself, being the combustion).

god: It's amazing how nothing can be something, isn't it?

MAX GOLD: It is.

(MAX GOLD cracks another Corona)

But it would be truly amazing to go back and see the very first time a conscious entity intentionally created something.

god: Indeed, but what really distinguishes your take on nothing and creation from the great Existentialists who equated nothingness with being, which leads to creation from nothing and, hence, god is no longer needed for there to be existence?

(MAX GOLD reaches into his pocket and pulls out a tablet of Ritalin. He crushes it in an empty beer cap and snorts it back. Shaking his head violently with his eyes rolled momentarily into the back of his head…)

MAX GOLD: Nothing doesn't own the franchise on the creation of all existence or existences. Energy is needed to create other forms of existence, above and beyond the primordial energy that nothing creates. The fact that nothingness created a substance and then that substance went through various mutations to create, or end up in, a conscious state or being – makes nothingness a philosophical creator and disproves the belief of an omnipotent creator. But, most forms of existence that are of any relevance to a conscious being require intent to create. Nothingness has no intent and can have no intent. And because nothingness has no intent, energy (God) is required because conscious energy (a part of God) is the only thing capable of making anything of any use to itself (conscious energy/God).

(Max Gold lights a cigarette)

Remember, when I refer to God, I am referring to all energy in existence and not a singular, conscious entity of some kind.

(god nods his head in approval)

The Existentialists are spot the fuck on with their main conclusion. However, they fail to grasp, or at least acknowledge, that energy (created by a completely unconscious and devoid thing like nothingness) has the innate ability to evolve to a point where it can consciously create existence, or things, independent of nothingness.

Nothingness and energy are not equal. However, all the energy in the universe is equal in that it's energy, regardless of the observable form of matter it may manifest in.

(taking a long drag then swigging from his beer)

There may be conscious beings in the universes that can actually create nothingness and use it… to create more energy (a process that could create infinity)! Nothingness can't do that. Nothingness can't create like that. Nothingness doesn't have the ability to become conscious – because to have consciousness you have to have energy – and before you can have energy you need nothingness – and nothingness possesses neither energy nor consciousness.

(MAX GOLD picks up a neatly folded beach towel off the cabana and blows his nose with it)

god: But, still, everything in existence was initially born from nothing.

MAX GOLD: 4.54 Billion years ago the earth was a stew of gasses, right?

god: Yes, essentially.

MAX GOLD: So was it nothing that created the existence of the current landscape on this planet?

god: *(smirking)* I see where you're headed.

MAX GOLD: Energy evolves by feeling out what patterns, system and combinations grow best to form something. This planet, and the things on it have come to be through the creative process of evolution; a process and existence which nothingness cannot take credit for. I just think it's important to focus on the fact that God is energy, we are energy and that, as such, we are a part of God. That said, Nothingness did start the first thing that ever was and some people need to get over the fact that there isn't a self-created omnipotent entity, or personal power, that's manning the controls for us and everything that ever was.

(god picks a toenail and flicks it into the sand)

People here just can't seem to get that the concept that consciousness can't create itself because wanting to create something requires intent, which requires consciousness... and consciousness cannot precede unconsciousness. Ergo, nothing gives birth to energy and energy gives birth to consciousness; we are, in essence, created by two things.

god: But why do you think humanity's understanding of reality makes it difficult to understand all of this? What is your reality?

MAX GOLD: *(taking a swig)* Reality is anything perceived by consciousness. Consciousness is energy comprised in such a way, usually manifest in the form of a brain, which enables consciousness to compute both internal and external stimuli, i.e., perceptions. Consciousness can be measured as intelligence; the higher the consciousness, the higher the intelligence.

What actually comprises reality is simply energy in different states, shapes and forms; not all of which are readily perceived by consciousness.

In essence, reality is energy perceiving itself.

What complicates the understanding of this is the human ego; the sense of self. The ego is a mechanism of the human psyche that provides the motivation for survival and pleasure.

To truly understand how the ego is created and how it dies is to understand the temporal significance of life.

The ego comes into existence through the process of physical maturation and experience. It's core is the body and face it identifies itself with and then the people and surroundings it bases its experiences on. An ego is as unique as a fingerprint. It never existed before the brain that houses it was formed and cannot exist without it.

(MAX GOLD lights a cigarette)

The human concept of an afterlife... sonofabitch!

THE LAST TEMPTATION of MAX GOLD

NIGHT III

(MAX GOLD inadvertently lit the filter of the cigarette. He lights another one properly and continues)

the human concept of an afterlife is nothing more than a coping mechanism of the ego, to aid it in its quest for survival. The modern ego, as guided by conventional wisdom, continuously fails to recognize that it is simply a form of energy (a part of reality) and that it will one day return to an unconscious form of energy.

Most people shudder to contemplate this fact. Many more simply refuse to accept it as fact.

Many philosophers have known this to be fact. Some have laughed at the absurdity of life as a result.

However, what can only be absurd is how we may, or may not, live life.

Life is an opportunity to experience the here and now. It is a gift both here and now. To focus or imagine an afterlife or, almost as foolishly, another human life after this one, is to waste what can be done and experienced with this life.

god: You just dismissed the concept of heaven and a core tenant of Buddhism is one fell bitch slap there.

MAX GOLD: I did, and it can't be done with enough force.

(MAX GOLD gets up to grab another Corona, opens it and takes a piss beside the lobster crate. Lighting up another cigarette he continues)

From day 1, it's been about making our desires a reality. These desires are rooted in alleviating pain; pain being the symptom of the need to meet survival. The weak philosophies that motivate weak minded people resolve that solutions to our pain and desires lie in another life; whether it's heaven or avoiding incarnation.

(MAX GOLD belches)

Fuck that. The reality I want to see is one where humans... end this stupid thing we call pain... here... on earth.

It's possible... and higher consciousness will make it a reality.

The very fact that once we die... that's it, it's over... is a tragedy second to the fact that our life can be, as it is for many, a tragedy in its existence.

While trying to contemplate other realities and possibilities that are different from the ones we experience in this one can lead to discoveries and improvement in this reality... it misses the point that the whole point of this reality is to focus on this one! This reality is the only thing that matters because we'll never know of a reality before it or after it... until we evolve to a point where such a case could even be the case... in this reality!

God: But how do people attain this higher consciousness?

MAX GOLD: There is no one way. There are just ways that are better for some than others. Communication for starters.

god: In what way?

MAX GOLD: People need to express themselves more and listen to themselves in the process. And people need to listen to people and learn and not feel so fucking weird, or threatened, in the process. And we all need to speak a common language.

(MAX GOLD throws an empty Corona under a Palm tree and opens another)

You know, some college kid reviewed Don't Judge a Dick by Its Foreskin and balked at my suggestion of the world adopting English as a common language. The kid didn't get that I don't care about the superiority of the English language. It's the largest second language in the world and likely the easiest way for 7 Billion to really start communicating with each other. Languages are barrier between people. Not being able to communicate with a person causes misunderstanding, fear, ignorance… the list of negatives is a fucking myriad god.

god: The Tower of Babel.

MAX GOLD: Fucking precisely! Different languages have the exact opposite effect of what a language's sole purpose is – to facilitate communication, i.e., accurate expression and understanding. If every person on the planet could talk, listen and relate to one another on the same page… think of the advancements we'd have in all of our pursuits?

(MAX GOLD plays in a pocket to find a match, but realizes he has run out. god gives him a lighter)

Thanks.

(MAX GOLD lights his cigarette)

I mean, holy shit, it really pisses me off that I can't sit down with the average person anywhere East of New York or West of Los Angeles and just shoot the shit about life. It troubles me that I can't listen to their news, or read their books and magazines and it's just as bad that they can't listen to and read what I can over here. There's a disconnect. We need to connect it; and not just through some fucking translator software - I mean on a person to person level. And a common language is the answer. Make it Russian, Mandarin, Swahili, whatever the fuck… we just need to be able to communicate like a global village. We should rename the English language to "World" or "Global".

god: Well, you nailed that one on the head

MAX GOLD: (scratching his head) Speaking of nailing things in the head, I thought I was meeting up with a snatch any time now. Fuck me.

Grow your own MUSHROOMS in your own HOME

MAGIC MUSHROOM TRAYS

Enjoy the thrill of picking fresh, delicious mushrooms at home. Keep your table supplied for months. The Magic Trays are completely developed with spawn. Simply place trays in basement, keep soil damp, mushrooms appear in two or three weeks. Each tray produces for two to four months. Simple instructions with each order. Magic Mushroom Trays are prepared by mushroom growers with years of experience, the same way as those used for commercial production. Trays are 14 x 18 inches. Shipping weight 25 lbs. Orders accepted all states—except West Coast and deep South. **Mail** order now. Enclose check or money order. 4 for $5.00—10 for $10. Trays sent Motor Express or Railway Express.

HORLOCHER MUSHROOM FARMS
ASHTABULA, OHIO

DAY IIII

god: But tell me what else could people do to expand consciousness?

MAX GOLD: Read and learn about psychology and psychiatry. Devour that shit.

god: *(in a suggestive manner)* Yes?

MAX GOLD: Read and learn about The Hero's Journey and anything by Joseph Campbell.

god: And?

MAX GOLD: Have a hallucinogenic experience. And don't be a fucking asshole about it and down a bunch a mushrooms like you would a bunch of beers. They can open your mind to the possibility that there are other possibilities out there. They're magic.

god: Anything else?

MAX GOLD: In the future we'll need longer lifespans facilitated by genetic alterations. And we'll need implants that are capable of hardwiring information to, and into, our senses and brains.

god: When you speak of humans having implants in their bodies that enable them to fuse all the collective knowledge known to the species into their brains and respective consciousness… that's artificial intelligence, I mean you'd be part machine?

MAX GOLD: You've made two assumptions. To answer your last assumption first, we already are a machine of sorts. We've taken the attributes from our organic environments and human physiology and applied them to more efficient man made things; things we call machines. Machines are a part of us, for arguments sake, in that we created them based on an ideal image of ourselves and our desires.

(MAX GOLD takes a deliberate drag on his cigarette)

By making, say, a computer chip that can be implanted in our brains to provide access to and literally infuse information into our brains… it's no different than using a piece of thread to stitch broken skin together… taking a pill to kill bacteria in our blood… transplanting a heart grown in a petri dish… altering genetic sequences to alter cell regeneration – it's just hyper evolved, relatively speaking. It boils down to making an improvement to our wellbeing that we inherently lack.

(MAX GOLD belches then hiccups)

But unlike stitches, meds, transplants and what not, a person infused with implanted intelligence might be able to pass that intelligence on to their offspring, in time…

(MAX GOLD takes a deep breath)

Like our fear of fucking snakes.

(MAX GOLD squints)

And if not, fuck it, just implant each new generation.

(MAX FOLD rips a wet fart)

god: *(in disgust)* Oh my fucking me, you were born in a barn.

MAX GOLD: So was Jesus.

(MAX GOLD throws his empty beer into the ocean and opens another while lighting a cigarette)

But to answer your first assumption, most intelligence is artificial.

god: I'm sorry, I don't follow.

MAX GOLD: Google the definition of artificial.

(MAX GOLD hands god his Apple iPhone™ and god Google's "artificial")

god: Shit, you're right...

> **Showing results for *artificial***
> Search instead for artifical
>
> # ar·ti·fi·cial
> /ˌärtəˈfiSHəl/
>
> *adjective*
>
> 1. made or produced by human beings rather than occurring naturally, typically as a copy of something natural.
>   "her skin glowed in the artificial light"
>
> 2. (of a person or a person's behavior) insincere or affected.
>   "an artificial smile"
>   *synonyms:* **insincere, feigned, false,** unnatural, contrived, put-on, exaggerated, **forced, labored,** strained, hollow; More

MAX GOLD: Of course. All of the intelligence we have now is artificial, but it's natural because we create it. Our knowledge base has progressed over thousands of years by individuals taking the knowledge they've had shared with them and the knowledge they've discovered, and vice versa, to then add to the knowledge base. Every kid born today is the recipient of intelligence that would have been inconceivable 5,000 years ago, let alone 100 years ago.

Think about it this way... none of us can make anything we use in a modern society. Not one person can manufacture a single fucking item needed. No one can mine a resource, make the machines needed to mine and transport the resources and then manufacture the resources into a usable thing.

(MAX GOLD pauses to offer god a Corona. god politely declines)

The process of manufacturing is an example of artificial intelligence. So if we make ourselves smarter with technology developed by a bunch of a smart people... who cares... we're all in it together and we'll benefit in the process. And it will be quite natural and organic in the process.

god: What is your stance on giving machines higher consciousness?

MAX GOLD: We're not ready to be like the idea of a god yet. No human is evolved enough to be able to weigh the benefits and risks of that kind of scenario. You don't want to give something inherently more efficient than yourself, equal or more capability than yourself. But who knows what transhuman intelligence will make possible. You have to stay open to new possibilities.

god: Is that all?

MAX GOLD: *(while moaning over a long boozy piss)* Reading the bibliography in "Don't Judge a Dick by Its Foreskin" wouldn't hurt my summation.

(MAX GOLD flicks his penis around in a helicopter blade motion as he finishes relieving himself. While reaching for another Corona, MAX GOLD trips on his shoelace and stumbles, knocking over a table and breaking an umbrella stand and the cabana itself)

NIGHT IIII

god: *(staring down at the sand)* Well Max, you've convinced me.

(god continues staring at the sand)

You've convinced me that there is hope for your species.

(god looks up)

The question is can you convince humanity of as much? Can humanity be convinced of as much?

MAX GOLD: Well, I don't want to convince anyone of anything.

god: Well then what do you want?

MAX GOLD: I want people to realize and choose what they want.

god: But how can people choose something without it being presented to them?

MAX GOLD: Simple.

god: How simple?

MAX GOLD: By putting ideas, options if you will, out there for people to embrace if they'd like, while giving them the choice to discard any ideas they don't like; free of fear, shame and judgment in the process.

(MAX GOLD lights a cigarette)

See, if you have to convince a person of something it means you have to overcome an obstacle they have to understanding or accepting what it is you're trying to convey. That's a situation that's tense in and of itself; you're trying to sell something.

(MAX GOLD pauses to pop a zit on his shoulder and then sips from his Corona)

If something makes sense... if it's true, people will get it. Many might not initially... if it's foreign to them or requires too much thought for them at first; but in time they'll understand, if it's true.

(MAX GOLD, while seated, urinates on a sandcastle made with his feet)

People need to find their own truth, god. And when they do, on their terms, they'll find love, understanding, compassion, fun and... themselves.

(MAX GOLD takes a last drag of his cigarette)

god: But what makes what you're talking about... true?

MAX GOLD: If, when you're alone, an idea makes sense in both your head and heart... and no one can be harmed because of it... it is true.

god: I'm sold.

MAX GOLD: Of course you are. You took the iPhone 666 words ago.

~

Night becomes day. Stillness returns to motion. Fear becomes hope.

~